ESPIONAGE BLACK BOOK FOUR

In this series:

ESPIONAGE BLACK BOOK FOUR:
Open-Source Intelligence Explained

Henry W. Prunckun

Bibliologica Press

Espionage Black Book Four:
Open-Source Intelligence Explained

Copyright © 2021 by Henry W. Prunckun

ISBN 978-0-6452362-1-7

A catalogue record for this
book is available from the
NATIONAL LIBRARY OF AUSTRALIA
National Library of Australia

For information on all Bibliologica Press's publications,
visit our Web site at bibliologica.com

Bibliologica Press
P.O. Box 656
Unley, South Australia, 5061
Australia

CONTENTS

— CHAPTER ONE —

OPEN-SOURCE INTELLIGENCE

I f *intelligence* is information that has been evaluated, then *open-source intelligence* (OSINT) is information from open sources that have been evaluated. The evaluation process involves a cyclical procedure that starts with defining the research question or making a statement to guide the inquiry. Arguably, this is the most critical aspect of secret research because without "compass bearing," an investigation can drift off course. Without proper orientation, decision-makers will not receive the advice they need.

The next step is to identify the information that can answer the question. Because intelligence research is conducted in secret, we might think that only secret information is used. Yet, there is no truth to this thought. The world is awash with data. It is all around us, and any piece of data in the public domain is free to gather.

What makes intelligence research different to other forms of research, is that some aspect of the process is secret. This could be the source, the way the data were collected, the analysis method, the weighting given to the conclusions, or the final report.

As an example, take the U.S. National Intelligence Estimates (NIE).[1] These reports are the product of the U.S. intelligence community's collective long-term

---

1. James R. Clapper with Trey Brown, *Facts and Fears: Hard Truths from a Life in Intelligence* (New York: Viking, 2018), pp. 71–72.

thinking on a particular issue. Although classified, these reports are, from time to time, released to the public. But before they are, the aspects of them that make them secret are removed.

Once collected, the evaluation process begins. Each data item needs to be assessed as to its accuracy and reliability. This is done the same way that a scholar evaluates information used in an academic study.[2] As a guide, intelligence analysts use the so-called *Admiralty Code* or *NATO System*. These systems are based on the *Admiralty Grading System* developed during World War II for tactical military purposes. At the time, the intention was to assign some level of certainty to information used in combat intelligence reports. It was an effective system and continued. We will be revisited in Chapter Six when we discuss validating information sources.

The analytic stage of the process is the part where the raw data is assessed, transforming it into intelligence. In Chapter Seven, we look at how this is done.

Once analyzed, the findings are compiled into report form and circulated to those with need-to-know or a right-to-know.[3] In intelligence terminology, this is called *dissemination* to intelligence *consumers*.

---

2. Henry W. Prunckun, *Writing a Criminal Justice Thesis* (Unley, South Australia: Bibliologica Press, 2021).

3. Need-to-know is tied to the doctrine of need-to-share, which states that it is important to share information across the intelligence community as well as within agencies. In determining the security needs of information, the doctrine reminds us that it is not absolute security that is sought, but a level of security in line with the sensitivity level of the information being guarded. And, the requirements of security should not eclipse the objectives of the secret research being conducted by impinge upon wider operational effectiveness.

These steps are called the *intelligence cycle*. Although practitioners and scholars talk about this cycle in slightly different terms, the underlining steps that move an analyst's thinking from question to answer are the same. Figure 1 shows one such view.

Figure 1—A typical view of the intelligence cycle.

HISTORY OF OPEN-SOURCE INTELLIGENCE

Arguably, open-source information was not the fountain of intelligence of early militaries, and hence political planning. Sun Tzu wrote, "Knowledge of the enemy's dispositions can only be obtained from other men."[4] The *Bible* cites gathering information by deploying spies.[5]

Information security should not inhibit the flow of information or hamper the dissemination of intelligence to a range of users.

4. Sun-tzu, translated by Samuel B. Griffith, *The Art of War* (Oxford: Clarendon Press, 1964).

5. *Numbers* 13: 14.

This is understandable because the sources of information were limited to what *a person* could collect by the five senses.

The coming of age of open-source information can be attributed to the Second World War when William "Wild Bill" Donovan posited that intelligence is neither mysterious nor sinister.[6] Agents of the Office of Strategic Services could obtain more helpful information in a few minutes spent with a freight train brakeman than Mata-Hari could in an entire evening.[7]

A case often cited as the quintessential instance of the start of open-source intelligence involved what was then the U.S. Foreign Broadcast Information Service.[8] The FBIS was created in 1941 to collect information that supported combat operations.[9]

---

6. William J. Donovan, "Intelligence: Key to Defense," in *Life*, September 30, 1946, p. 114.

7. Harry Howe Ransom, *The Intelligence Establishment* (Cambridge MA: Harvard University Press, 1970), p 17.

8. Despite a search of the literature, a reliable reference to this repeatedly told story could not be found. Although there is a mention a government source on a U.S. Army Web page, but this on-line article did not cite its source. Nevertheless, even if this story is one day to found to lack a factual base, it can still serve as a notional example of how effective open-source information can be. See: www.army.mil/article/94007/Service _members__civilians_learn_to_harness_power_of__Open_So urce__information (accessed July 16, 2021).

9. The FBIS operated from 1941 to 1996 when it was absorbed into the Central Intelligence Agency. FBIS translated newspapers, periodicals, government announcements, as well as translated radio, and then added television, broadcasts to the list of open sources that targeted countries of interest. Kalev Leetaru, "The Scope of FBIS and BBC Open-Source Media

The often-cited example centers on Allied forces' need to know how effective their aerial bombing campaign was in the days before the D-Day invasion. Specifically, they wanted to know if the railway bridges in Occupied France had been disabled. This assessment was necessary because the Allies wanted to make it as difficult as possible for Nazi forces to send troops and supplies once the invasion started.

If Hollywood was telling the story, it might lead us to believe that this was done through the derring-do exploits of an elite unit of covert operatives parachuted behind enemy lines. On the ground, these Hollywood operatives would stealthily traverse the country to carry out surveillance of every bombed bridge, radioing damage assessments as their mission progressed.

The plot would make for good viewing, but such a story would not only be fictitious but fanciful. The weaknesses of this type of data collection plan do not have to be spelled out to be understood.

Instead, the Allies relied on information provided by those living in Occupied France. These data were provided daily, unencoded, and in the open. Without realizing it, these data became a rich source of information for Allied intelligence.

What was this source? What were these data? It was the daily radio news broadcasts to the local population regarding the price of farm produce, in particular, the cost of oranges in Paris.

These market reports were monitored by FBIS. Its linguists translated much of the shortwave transmissions

---

Coverage, 1979–2008," in *Studies in Intelligence*, Volume 54, Number 1, pp. 17–37.

from Occupied Europe into English for use by the intelligence community.

So, what does the price of oranges have to do with bridge damage assessments? It is what social scientists call *unobtrusive data*.[10] That is, rather than relying on the reports of undercover agents, covert operators, or commandos, analysts observed actual events rather than reposted events. This underscores some of the advantages pointed out some 50 years later: "The study of *actual* rather than *reported* behaviour; safety; repeatability; non-disruptive, non-reactive; easy accessibility; inexpensive; and good source of longitudinal data."[11]

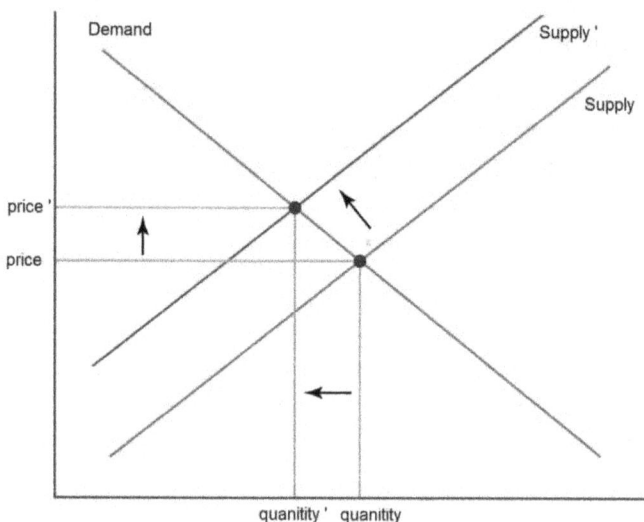

Figure 2—Reduction in supply results in increased price.

10. Allen Kellehear, *The Unobtrusive Researcher: A Guide to Methods* (St. Leonards, NSW: Allen & Unwin, 1993).

11. Kellehear, *The Unobtrusive Researcher: A Guide to Methods* , p. 7.

To do this, analysts used the price of oranges as the metric for battle damage. As a fruit that was imported from North Africa (i.e., not regionally grown), analysts were able to measure the effects of the bombing campaign (i.e., the inability of trains to transport oranges to Paris), rather than reported conduct, which was fair to say difficult and perhaps impractical to collect. And, if were to be achieved, subject to reporting inaccuracies.[12]

Using the microeconomic theory of supply-and-demand, Allied intelligence reasoned that because demand remained constant, and the price of oranges increased, this was due to diminished supply (see Figure 2). Because the only method of shipping oranges to Paris was via the rail network, Allied intelligence concluded that the air campaign was effective in cutting off orange shipments, and as such, would also cut off Nazi troop transport and military supplies.

In sum, "...the great benefit of unobtrusive measures lies in their 'non-reactive' character, thus avoiding threats to data quality caused by the [social science] researcher's presence."[13] Because intelligence research is based on qualitative and quantitative social science research methods,[14] then it follows that the same can be said for

---

12. Social scientists have known for over half-a-century that the act of asking someone a question—in this case, an agent, covert operative, etc.—may unintentionally influence the "character" of the answer. Raymond M. Lee, "Unobtrusive Methods, History of," in James D. Wright, editor, *International Encyclopedia of the Social & Behavioral Sciences, 2nd edition*, Volume 24, 2015, p. 767.

13. Lee, "Unobtrusive Methods, History of," p. 767.

14. Hank Prunckun, *Methods of Inquiry for Intelligence Analysis, Third Edition* (Lanham, MD: Rowman & Littlefield, 2019), p. 9.

secret research. Although in intelligence research, the term that is used is *open-source* information, which is then analyzed to produce open-source intelligence.

Open-source intelligence refers to a methodology for collecting data in the public domain, but it is used for intelligence research. This approach contrasts to covert and clandestine methods of data collection, which are most often associated with intelligence work.

Today, OSINT is universally used by military and national security agencies, law enforcement organizations, and businesses to produce competitive intelligence. "Freely available, unclassified materials provide contexts for all kinds of other reporting. In other words, all analysts are—and must be—all-source analysts."[15]

Open-source intelligence is also used by non-government organizations such as Human Rights Watch and Amnesty International to gather facts about contraventions to international law. For example, "In June 2014, Human Rights Watch used satellite imagery combined with videos posted online by the Islamic State to find the exact location of mass executions in Tikrit, Iraq."[16]

Although "…open-source information probably will never provide the 'smoking gun' about some issue or

---

15. Thomas Fingar, "A Guide to All-Source Analysis," in Peter C. Oleson, editor, *AFIO's Guide to the Study of Intelligence* (Falls Church, VA: Association of Former Intelligence Officers, 2016), pp. 297–298.

16. Eliot Higgins, "A New Age of Open-Source Investigation: International Examples," in Babak Akhgar, P. Saskia Bayerl, and Fraser Sampson, Editors, *Open-Source Intelligence Investigation: From Strategy to Implementation* (Cham, Switzerland: Springer International Publishing, 2016), p. 190.

threat, but it can be instrumental in helping analysts to focus better or 'drive' clandestine collection activities by first identifying what is truly secret."[17]

There are many sources of information—open-source just being one. So, it warrants an examination of the range of data available by looking at the taxonomy of intelligence sources.

---

17. Richard A. Best Jr. and Alfred Cumming, *Open-Source Intelligence: Issues for Congress* (Washington, DC: Congressional Research Service, 2008), p. 3.

— CHAPTER TWO —

TAXONOMY OF INTELLIGENCE SOURCES

B y definition, open-source intelligence is publicly available information. As such, there is little to limit—either ethically or legally—analysts accessing these data.[18] These sources are in direct contrast to covert and clandestine methods, arguably the methods commonly associated with intelligence work.

In the parlance of social science research, open-source information is categorized as secondary data, and covert and clandestine information can be likened to primary data.

Primary data are information collected by the intelligence researcher (or their agents) for a specific project. For instance, an analyst may identify in their information collection plan that they need photographs of a secluded bridge over the Orrenabad River. In this case, an operative may be sent to the location and, perhaps, using a cover story, would take photographs.

---

18. Although for some agencies, in particular the military and national security agencies, there may be restriction imposed by regulations or directives that prohibit the collection, retention, or dissemination of information regarding U.S. citizens. See, for instance, Army Regulation 381–10, *U.S. Army Intelligence Activities*, and Executive Order 12333, *U.S. Intelligence Activities*. The ethical issues regarding the collection and retention of open-source intelligence will be discussed in Chapter Nine.

In contrast, secondary data are collected for another purpose (and by others) but can be applied to the question under investigation. Again, using the Orrenabad River example, secondary data may include commercially available photographs of the bridge from satellites; or photographs of the bridge taken by recent travelers who have posted these on their social media websites; or photographs that related to tourist promotional material (on the Web or in printed form in libraries); or any number of other publicly available sources.

The spectrum of information sources ranges from open and semi-open sources to clandestine and covert. Having started our examination of open-source information, we need to discuss sources at the other end of the information spectrum—those considered surreptitious in some way.

Because of the intrinsically safe nature of open-source information, the same high level of consideration and planning is not required to obtain these data compared to organizing data collection by clandestine and covert means.

Although akin to covert collection, clandestine data collection is different. Clandestine operations operate in the open—visible to the target but disguised so that they do not appear to be what they seem. Covert operations, in contrast, are carried out in secret. They are hidden and not visible to the target even in a disguised form; they are intrusive, but because they are invisible, the target does not know that an operation is being conducted.[19]

Covert methods include undercover operatives, informants/agents, physical surveillance, electronic

---

19. The British refer to covert operations as *special operations*. Frederick P. Hitz, *The Great Game: The Myth and Reality of Espionage* (New York: Alfred A. Knoff, 2004), p. 5.

surveillance, informants, mail covers, and waste collection. Analysts may want to use these methods in their collection plans because an attempt to obtain information via open methods might be met with a nil result. For instance, secretive methods are used where the target is concealing information. The only way to obtain such data under these circumstances might be to penetrate the security measures via one of these two surreptitious means.

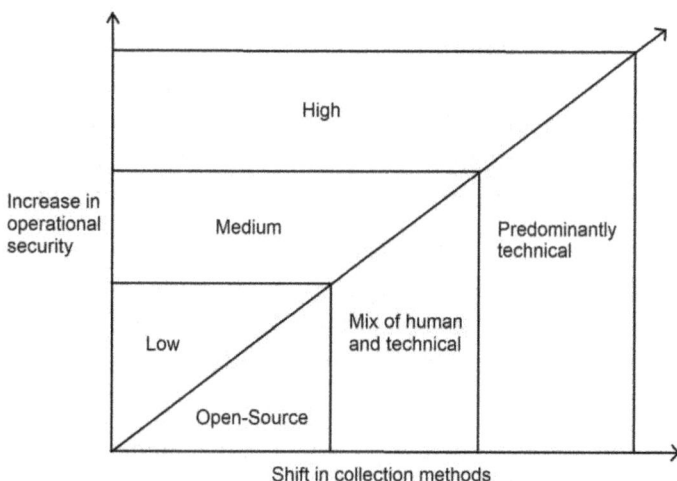

Figure 3—From open-source information to highly technical secret collection.[20]

As a principle, when the opposition uses operational security measures (i.e., defensive counterintelligence), these measures will prevent—or severely limit—an agent's ability to gather information. But security is not the only reason. Other factors that inhibit collection include an agent's cultural, ethnic, religious, language, or

---

20. Source: Hank Prunckun, *Scientific Methods of Inquiry for Intelligence Analysis, Third Edition* (Lanham, MD: Rowman & Littlefield, 2019), p. 63

other barriers. For instance, an agent's ethnic background may be prevented from penetrating, say, a right-wing political group or an outlaw motorcycle gang. Data collection methods will shift to more technical means, such as electronic, optical, and physical surveillance in such situations.[21] The relationship between a target's level of operational security and an analyst's need to shift to technical collection methods is shown in Figure 3.

Gauging from Figure 3, we might be tempted to conclude that most intelligence research material comes from technical collections, given that two-thirds of the area under the graph relates to secret methods. However, as Professor Harry Howe Ransom wrote in his influential work on intelligence: "...95 per cent of peacetime intelligence [comes] from open sources."[22] Professor Ransom's analysis of the United States's national intelligence collection effort stated that more than 80 percent came from "overt, above-board methods [that] would normally be available to anyone with a well-organized information gathering system."[23]

The late Richard Helms, former director of central intelligence, stated that between the end of the Second World War when the Office of Strategic Services was

---

21.   Henry Prunckun, *Espionage Black Book Three: Surveillance Explained* (Unley, South Australia: Bibliologica Press, 2021).

22.   Harry Howe Ransom, *The Intelligence Establishment* (Cambridge, MA: Harvard University Press, 1971), p. 19. Professor Ransom was quoting Ellis M. Zacharias, a World War II deputy director of the Office of Naval Intelligence. According to Zacharias, only four percent of intelligence came from semi-open sources, and a mere one percent from secret agents.

23. Ransom, *The Intelligence Establishment*, p. 20.

deactivated and 1947 when the CIA was created, the agency responsible for secret intelligence was the Strategic Services Unit (SSU). In conducting its intelligence research on the Union of the Soviet Socialist Republics, the SSU used the Library of Congress as its primary data source.[24] All the Library of Congress's data about the USSR were publicly available.

Regarding business intelligence, it has been estimated that "...90 percent of all information that you and your business need to make key decisions and to understand your market and competitors is already public or can be systematically developed from public data."[25] As an illustrative point from history, take this Cold War example. Polish intelligence officer Colonel Pawel Monat, a military attaché in Washington, DC, saved his Communist government large sums of money, time, and effort by accessing open-source information about commercial aviation "secrets." Regarding one experience in particular—that is, with *Aviation Week* magazine—he wrote: "Very little of this information was of really classified nature. We could have dug up most of it ourselves from other sources. But it would have taken us months of work and required us to shell out thousands of dollars to various agents to ferret out the facts, one by one. The magazine handed it all to us on a silver platter [for fifty cents]."[26]

---

24. Richard Helms with William Hood, *A Look Over My Shoulder: A Life in the Central Intelligence Agency* (New York: Random House, 2003), p. 73.

25. John J. McGonagle Jr. and Carolyn M. Vella, *Outsmarting the Competition: Practical Approaches to Finding and Using Competitive Information* (Naperville, IL: Sourcebooks, 1990), p. 4.

26. Pawel Monat with John Dille, *Spy in the U.S.* (London: Frederick Muller Limited, 1962), p. 120.

More recently, the late Tom Clancy was queried about his infallible knowledge of some obscure technical and scientific details in his espionage novels. He is reported to have denied having access to classified defense information but instead pointed out what others have discovered—it could all be found in the open-source literature.[27]   And, intelligence scholar Loch Johnson supports Clancy's claim: "The overwhelming percentage of information in intelligence reports comes from open-source searches, augmented by a small percentage of clandestinely derived data."[28]

These views on the usefulness of open-source data are also reflected in the social sciences.

> C. Wright Mills, among others, always argued that you should never do any empirical research unless you absolutely had to. ... He was of the view that we had more than enough data on a whole range of human issues, and that the central problem confronting modern social science was that of interpreting it all.[29]

---

27. Frederick P. Hitz, *The Great Game: The Myth and Reality of Espionage* (New York: Alfred A. Knopf, 2004), p. 86.

28. Loch K. Johnson, "Sketches for a Theory of Strategic Intelligence," in Peter Gill, Stephen Marrin, and Mark Phythian, editors, *Intelligence Theory: Key Questions and Debates* (London: Routledge, 2009), p. 42.

29. Kellehear, *The Unobtrusive Researcher: A Guide to Methods* , p. 53.

— CHAPTER THREE —

INTELLIGENCE COMMUNITY

I s open-source intelligence better suited to the needs of particular agencies? The short answer is, "no." Arguably, all intelligence agencies use open sources in their production of intelligence reports. Because the names, roles, and lines of responsibilities of individual intelligence agencies shift with the rhythm of their elected political overseers,[30] it will be more beneficial to survey the structure of intelligence, citing examples rather than presenting a comprehensive list of the actors. Or, put another way, we will discuss the typology of the intelligence community (IC).[31] In this way, we can appreciate how the intelligence community uses open-source information.

Intelligence can be classified into five types. Although each type of intelligence has a unique purpose, they are in many ways related. Notably, the same methods of operation, tactics, devices, information storage systems, and analytic methods are used by each. In addition, information itself holds no bounds as to its usefulness, and a particular piece of information could be the target for more than one type of intelligence user. In other words, the primary

---

30. Usually after a political development, followed by a review.

31. This discussion has been adapted from Henry Prunckun, *Information Security: A Practical Handbook on Business Counterintelligence* (South Australia: Bibliologica Press, 2020), pp. 3–5.

difference between the various types of intelligence lies in their end purpose or intent.

The five types of intelligence will be classified as national security intelligence, military intelligence, law enforcement intelligence, business intelligence,and private intelligence for our discussion.[32]

National security intelligence relates to secret research conducted by the various branches of a nation's foreign diplomatic service and overseas intelligence agencies. Depending on the country, it could also include its atomic energy authority. Western nations generally tend to have a central agency that coordinates the numerous intelligence functions, organizations, and the collection and processing of information.[33] Russia[34] and China,[35] in contrast, lean toward a unified system with one supreme agency taking on all three roles—collection, analysis, and coordination.

---

32. The author's typology parallels a similar "seven tribes" categorization developed by Steele, who posited that there are seven classifications: (1) government, (2) military, (3) law enforcement, (4) business, (5) academia, (6) nongovernmental organizations and the media, and (7) citizen advocacy groups, labor unions, and religious organizations. Robert David Steele, "Open-Source Intelligence," in Loch K. Johnson, *Strategic Intelligence: Understanding the Hidden Side of Government, Volume 1* (Westport, CT: Praeger, 2007), pp. 95 and 116fn1.

33. Patrick F. Walsh, *Intelligence and Intelligence Analysis* (London: Routledge, 2011).

34. Eberhard Schneider, "The Russian Federal Security Service under President Putin," in Stephen White, editor, *Politics and the Ruling Group in Putin's Russia. Studies in Central and Eastern Europe* (London: Palgrave, 2008), pp. 42–62.

35. Peter Mattis and Matthew Brazil, *Chinese Communist Espionage: An Intelligence Primer* (Annapolis, MD: Naval Institute Press, 2019).

The types of information sought by national security intelligence analysts can be anything from the current political issues facing foreign governments, the health, education, and social structures of countries of interest, their social problems, and their legal institutions.

Analysts may also seek information concerning the availability of natural resources, international trade relationships, and the state of the global monetary order. Without a doubt, they would also look for information on foreign technological developments, nuclear matters, and almost anything to do with foreign defense industries.

These agencies operate under names that suggest their interest is in the areas of "media monitoring." For instance, the former U.S. Foreign Broadcast Information Service operated from 1941 to 2005 alongside the U.S. Defense Department's Joint Publications Research Service.[36]

These agencies monitored newspapers, magazines, trade publications, television and radio stations, news services, news tickers, digital platforms, and podcasts promoted by these outlets, as well as politicians, commentators, and social media influencers.[37]

---

36. As an example, the American Open-Source Center, later to become the Open-Source Enterprise, which incorporated the CIA's Directorate of Digital Innovation. There is also the Library of Congress's Federal Research Division that conducts customized open-source research for the executive branch. Some of the types of reports produced include annotated bibliographies, organizational and legislative histories, subject specific studies, and books like the series titled *Country Studies*. The latter addresses historical, social, economic, military, and political issues of countries around the world.

37. Princeton University was reported to have monitored foreign shortwave broadcasts in the 1930 before the FBIS took

Military intelligence examines the military forces of other countries—their force composition and deployment, doctrine, and plans, as well as other country's warfighting operations. Although information is gathered from numerous sources, open-source information features prominent. By way of example, the U.S. Defense Department has an OSINT program that comprises military offices such as the Defense Intelligence Agency, the National Geospatial-Intelligence Agency, and the Army Foreign Military Studies Office.

Law enforcement intelligence encompasses those agencies engaged in a defensive counterintelligence function.[38] They would include a nation's police and law enforcement, compliance and regulatory agencies (which can be numerous), immigration and customs services. And, as the case may be, specific agencies created to address the threat from foreign and internal subversion, espionage, sabotage, or terrorism (i.e., offensive counterintelligence).

The law enforcement agencies use OSINT to deter and prevent crime and investigate criminal activity. OSINT is a staple of many domestic intelligence Fusion Centers and large investigative agencies such as Scotland Yard, the Royal Canadian Mounted Police, INTERPOL, and EUROPOL, in addition to big-city police departments such as those in New York City and Los Angeles.

---

over. Stephen Mercado, "A Venerable Source in a New Era: Sailing the Sea of OSINT in the Information Age," in Christopher Andrew, Richard J. Aldrich, and Wesley K. Wark, editors, *Secret Intelligence: A Reader* (London: Routledge, 2009), p. 78.

38. Arthur E. Gerringer and Josh Bart, "Law Enforcement Intelligence," in Peter C. Oleson, editor, *AFIO's Guide to the Study of Intelligence* (Falls Church, VA: Association of Former Intelligence Officers, 2016), pp. 321–326.

Business intelligence is concerned with the acquisition of trade information and commercial data about competing firms. It is sometimes referred to as *competitor intelligence*, *commercial intelligence*, *corporate intelligence*, and in less polite language, *industrial espionage*. This type of secret research can occur on a local, regional, national, and international basis.

Business intelligence is not limited to businesses that carry out research themselves but can also include information brokers[39] and private investigation firms and in some countries such as Russia, China, and North Korea, by their national security/military intelligence units.[40]

The structure of private intelligence is diverse; therefore, this survey will limit its discussion to those firms and inquiry agents who offer their services and expertise in intelligence work to the public for a fee or other reward. Although the term *public* implies individuals, there is some overlap in what constitutes private intelligence, business intelligence, or even national security intelligence. The final determination of what category it falls into depends on who is hiring the agent or researcher.

Private intelligence practitioners offer a range of specialist services that go beyond the bounds of the average private investigator. Often, the private intelligence practitioner comes from a background in law enforcement, military, or national security intelligence work. Their specialties may be in background investigations or surveillance. They may have training in the use of optical

---

39. Information brokers are businesses that collect personal and corporate data from public records. They then sell or license access to their databases to third parties.

40 . By way of example, see Clive Hamilton, *Silent Invasion: China's Influence in Australia* (Melbourne: Hardie Grant Books, 2018), pp. 151–176.

or electronic audio surveillance equipment, and they would be familiar with intelligence analysis methods. These practitioners may offer advice on defensive business counterintelligence and electronic audio countermeasures (debugging). They may also specialize in personal security for VIPs (i.e., body-guarding).

For purposes of our discussion, private intelligence agencies include privately run organizations that maintain large databases for specialized inquiry work, for example, credit reporting agencies and information brokers. This type of intelligence work also incorporates private investigation firms.

OSINT is now viewed as an essential source of information because, as the authors of *The Commission on the Intelligence Capabilities of the United States Regarding Weapons of Mass Destruction* stated:

The ever-shifting nature of our intelligence needs compels the Intelligence Community to quickly and easily understand a wide range of foreign countries and cultures. As we have discussed, today's threats are rapidly changing and geographically diffuse; it is a fact of life that an intelligence analyst may be forced to shift rapidly from one topic to the next. Increasingly, Intelligence Community professionals need to quickly assimilate social, economic, and cultural information about a country—information often detailed in open sources.

Open-source information provides a base for understanding classified materials. Despite large quantities of classified material produced by the Intelligence Community, the amount of classified information produced on any one topic can be quite limited, and may be taken out of context if viewed only from a classified-source perspective. Perhaps the most important example today relates to terrorism, where open-source information can fill gaps and create links that allow analysts to better understand fragmented

21

intelligence, rumored terrorist plans, possible means of attack, and potential targets.

Open-source materials can protect sources and methods. Sometimes an intelligence judgment that is actually informed with sensitive, classified information can be defended on the basis of open-source reporting. This can prove useful when policymakers need to explain policy decisions or communicate with foreign officials without compromising classified sources.

Only open source can 'store history.'[41] A robust open-source program can, in effect, gather data to monitor the world's cultures and how they change with time. This is difficult, if not impossible, using the 'snap-shots' provided by classified collection methods.[42]

---

41. Known as *basic intelligence*. However, the term is a misnomer. "The term infers that somehow it is elementary or simple; but it is neither of these things. Basic intelligence is concerned with analyzing historical topics. The purpose is to provide information that can be used for a variety of research projects as well as operational reasons. A simple example of the latter use is where an operative or agent is developing a 'cover' or 'legend' and needs factual information about what a place looked like at a particular period in time." Hank Prunckun, *Method of Inquiry for Intelligence Analysis, Third Edition* (Lanham, MD: Rowman & Littlefiled, 2019), p. 15.

42. Laurence H. Silberman, et al., *Report to the President of the United States* (Washington, DC: The Commission on the Intelligence Capabilities of the United States Regarding Weapons of Mass Destruction, 2005), pp. 378–379.

— CHAPTER FOUR —

INFORMING INVESTIGATIONS

I t is one thing to discuss the types of information that comprise OSINT, but it is another to focus on how intelligence analysts can use open-source intelligence to support an investigation. Bertram pointed out that:

> OSINT is not the silver bullet that will blow your stalled investigation wide open, and neither is it just another faddy 'must have' capability; it is certainly not a frightening and soulless computer-based technology that is here to take your job. Instead, OSINT is just another arrow within the quiver of the investigative analyst, just like techniques such as interviewing, surveillance, fingerprinting and any number of others open to the skilled professional investigator or analyst.[43]

Let us start at the beginning of an intelligence project. Imagine that a state or provincial police department is investigating an issue of rebirthing of stolen vehicles. The detective branch may have been alerted to the problem via several methods, but regardless, the officer-in-charge has asked the branch's intelligence officer to compile a report.

As with any research project—one of general personal interest, an academic paper, or a secret intelligence report—the *research question*[44] is the keystone to what

---

43. Stewart Bertram, *The Tao of Open-Source Intelligence* (Ely, Cambridgeshire, UK: IT Governance Publishing, 2015), "Introduction."

44. Also known as a *statement of purpose.* John W. Creswell, *Essential Skills for the Qualitative Researcher* (Thousand Oaks, CA: Sage, 2016).

follows. This is because the research question sets the bearing as to where the inquiry goes. From this, the analyst can write what will be in-scope—that is, what will be included and what will be excluded by default. These need to be signed-off by the officer-in-charge to ensure that the project meets the branch's needs.

Let us take a concrete example to demonstrate the initial phase of an intelligence project. Returning to our vehicle rebirthing illustration, the request may be as vague as, "We have a problem in Hampden County. It looks like it's being used as a hub for transshipping vehicles to other parts of the country for rebirthing. What do we know about any of this?"

What do we know about any of this? is not a research question because it is too vague. The question needs to set the context so that it can be narrowed. There are several ways to narrow a research question—by time, location, actors, or another variable that allows specificity. In general, the narrower the problem is defined, the better the question.[45] Nevertheless, several variables can be used in constructing a research question. A question does not need to be restricted to a single factor.

What would be a better research question? Here is one to consider: "We have intelligence that indicates a criminal group has set-up in Hampden County. Reports suggest that this group has been operating to transship stolen vehicles to other parts of the country for the past three months. To focus the department's investigative resources, we need to know if this is an isolated business, or is it part of a larger organized criminal enterprise?"

---

45. For more on developing a research question, see Henry W. Prunckun, *Writing a Criminal Justice Thesis* (South Australia: Bibliologica Press, 2019).

Why is it a better question? Because it lays out the context of the problem so a specific request can be stated—*is this an isolated business, or is it part of a larger organized criminal enterprise?*

An analyst can then devise a data collection plan with this research question—i.e., what information will be needed to answer the question. Note that the plan only needs to specify the information that will help answers the research question. Any "nice to have" or "oh, that's fascinating" information should not form part of the plan. This triaging is the purpose of *scope*—the research question states what is within the project's scope and what is out of scope. Data items that invoke reactions such as, "That would be interesting to include," are out of scope.

In this hypothetical case, what might the information collection plan be looking for? Recall, we are only looking at open-source information here. In an actual situation, an analyst might be able to use internally generated criminal reports and access federal criminal databases to supplement their report, but for our purposes, we will restrict our gaze to open sources.

Since the research question is about an operational issue, the project lends itself to an *operational assessment*. Operational matters tend to be issues that are "...broader than what is happening at the tactical level (e.g., investigation and apprehension) but not prognostic as strategic intelligence would be (i.e., long-term implications)..."[46] Consequently, this type of report is well suited to the request we are considering. An example of a notional operational assessment appears in Chapter Seven. This example will help you visualize the product.

---

46. Prunckun, *Methods of Inquiry for Intelligence Analysis, Third Edition*, p. 154.

Brainstorming is an excellent method to explore the types of information needed and their possible sources.[47] In our case, we need to know about the enterprise of rebirthing so we can gauge whether the Hampden County situation is isolated or part of a more significant venture.

The list presented in Table 1 provides an idea of what data items are available for our question. This list is not meant to be exhaustive, just suggestive. Likewise, the sources will vary in name depending on the jurisdiction. Nevertheless, brainstorming works well to produce a collection plan.

Table 1 underscores that although open-source intelligence is not new, its advent has produced volumes of information that are easy to search. Previously, OSINT relied on paper-based resources found in libraries, newspaper archives, and the like. Although these physical sources are still applicable—especially where digitation has not taken place—the Internet has transformed OSINT.

Once the information needed to answer the research question has been collected and collated, the writing work beginning. It is usual for an analyst to produce several drafts before the final copy is sent up the chain of command. In addition to correcting grammar, diction, and syntax errors, the draft must not drift into areas outside the project's scope.

So, when editing, analysts should ask themselves whether the information in each section addresses the topic heading, ensuring there is no repeated information from elsewhere in the document. Moreover, each section needs to lead the reader to the research question and what that answer means for practice and/or policy.

---

47. A good text to complement brainstorming is Tony Buzan, *How to Mind Map* (London: HarperCollins, 2002).

Table 1—Indicative list of data items and their public sources.

Information Item	Source
Register of company and business names	Articles of incorporation and list of the business's principals
Lands title register	Details of the property's ownership
Google street view	Location coordinates and appearance of the vehicle warehouse
LinkedIn	Background and employment history of the business' principals
Facebook	Photographs of people of interest, as well as personal details and timelines of events
Newspapers/news reporting	Either paper copies or online, these can provide information on the principals of the business and such things as court appearances
Academic papers, master's theses, and doctoral dissertations	Background on the function, structure, and operating methods of outlaw motorcycle gangs in general, and the gang under investigation in particular

So far, we have discussed what an analyst can do when given a formal request. What about situations where an analyst is assigned to an investigative team under, say, the command of a police detective who has little or no training in intelligence? In these cases, the analyst may find themselves sitting at the outer edge of the circle with little to do.

The analyst needs to take the initiative by considering what they can do to help advance the investigation. Although we will be discussing a law enforcement operation, the same thinking applies to business intelligence, national security intelligence, and national security intelligence research projects.

The way to do this is to return to the *first principles* of intelligence—i.e., the most important reasons for conducting intelligence research. If intelligence reduces uncertainty in decision making and an investigation is about discovering facts, then why not assist detectives by compiling reports that provide insights about different aspects of the inquiry?

What might these reports look like? This first type that presents itself is the *target profile*.[48] Other types include biographical sketches, chronologies, and bulletins. If you were to skim a textbook on crime analysis techniques, more analytic methods could generate specific reports for investigation teams and managers.

The issue is that without knowing the research question, it is only possible to describe these methods and the types of data used. Nonetheless, it is possible to provide outlines of the sections these types of reports should contain to be useful. It is then up to the analyst to start the process of looking at what the investigative team is trying to do, what data is needed to answer the question,

---

48. Some law enforcement agencies use the terms *violator file summaries* and *target identification reports*. There may be other terms in circulation, but their purpose of this type of report is the same. See, Marilyn B. Peterson, editor, *Intelligence 2000: Revising the Basic Elements* (Lawrenceville, NJ: International Association of Law Enforcement Intelligence Analysts, 2000), p.127.

and what type of analysis can be applied to produce a report of value.[49]

In Chapter Eight, we will look closer at operational assessments, target profiles, and some of the other types of reports that can be produced. In the meantime, let us look at open-source information collection.

---

49. A useful book about analytic methods is Richards J. Heuer Jr. and Randolph H. Pherson, *Structured Analytic Techniques for Intelligence Analysis* (Washington, DC: CQ Press, 2011).

— CHAPTER FIVE —

COLLECTION

Open sources, used astutely, can be a boon to information collected by covert and clandestine means. Although, one should not replace the other.[50] Nevertheless, one of the concerns for covert data collection is safety to the officer or agent tasked with collecting information. The risks are real, and the consequences of failure are grave—both to the operative and the employing agency.[51]

However, open-source data collection does not pose the same risks, and the depth and breadth of the information are potentially vast. Having said that, the latter point can also be a drawback—often, if a specific piece of information is needed, it may be classified. Hence, an agent working undercover may be the only way to gather it. In this regard, volumes of information cannot be substituted for specificity.

Still, open-source information can quickly supply data to answer the research question and, therefore, assess a developing situation. It can also be used to determine the

---

50. The prevalence of open-source information should by no means underestimate the value of covert and clandestine data collection as well as data obtained from confidential sources. Robert Baer, *See No Evil: The True Story of a Ground Soldier in the CIA's War on Terrorism* (New York: Crown Publishers, 2002); and Melissa Boyle Mahle, *Denial and Deception: An Insider's View of the CIA from Iran-Contra to 9/11* (New York: Nation Books, 2004).

51. See Ted Gup, *The Book of Honor: Covert Lives and Classified Deaths at the CIA* (New York: Doubleday, 2000).

missing information (i.e., the *specificity* just discussed) by highlighting what is required in an information collection plan. Such a plan may, at this stage, point to the need to involve covert means.

A sometimes overlooked source of open-source information is the public and university libraries. Analysts should never forget the time-honored library.[52] This is a rich source of data.

Location-dependent, the quality and quantity of a library's holdings will vary. Still, generally, libraries have extensive holdings of nonfiction reference works as well as access to the interlibrary loan program. The latter provides access to nationwide holdings through member libraries.

In addition to the wealth of information in the library's stacks, libraries have reference books, maps, newspapers, journals, periodicals, registers, and catalogs. There are also special collections within libraries and privately maintained computer databases that contain topics of particular interest. The Russell J. Bowen Collection of Works on Intelligence, Security, and Covert Activities is an example of the former, housed in the Lauinger Library, Georgetown University, Washington, DC.

Before digital collections, searching and retrieval of information from a library was a time-consuming task. It sometimes required the analyst to search hard-copy index cards, cross-check the desired entry with another list of journals held by the library, and then locate and copy the article. If the library did not hold the journal, then the

---

52. Robert M. Clark, *Intelligence Analysis: A Target-Centric Approach, Second Edition* (Washington, DC: CQ Press, 2007), pp. 86–87.

process of acquiring it via internal loan or physically visiting another library would be the only option.

However, digital collections have eliminated these inefficiencies. Now an analyst can search a library that may not even be in their state or country from their computer workstation. The analyst can locate a journal article from that search and then download it in a readable format that looks exactly like the printed form in the hard-copy journal. Many books can also be downloaded in electronic form.[53]

Analysts can put together their special collection of intelligence reference books too. This can be done relatively inexpensively by visiting second-hand bookstores in their area. There are often titles relating to many areas of interest to scholar-spies—research methodologies, analysis, statistics, as well as a range of topics within intelligence and counterintelligence (e.g., cryptanalysis, espionage tradecraft, history of intelligence, spy memoirs, investigative exposés, etc.). Analysts can also buy books on subject areas that they might be focusing on in their investigations—arms control, weapons proliferation, transnational crime, terrorism, arms and drug trafficking, organized crime, corruption, cybercrime, war crimes—and the list could go on.

If analysts are located near a college or university, they might be able to find second-hand bookstores that have taken in last semester's student texts. These will have a potentially richer source of book titles. Although buying books this way is largely a hit-or-miss affair, it is an

---

53. As an aside, a growing number of analysts have degrees in library science. This skill set has commonalities for both professions. Personal communication with a former librarian at CIA, Dr. Edna Reid, Federal Bureau of Investigation, Washington, D.C., June 7, 2011.

inexpensive method of acquiring a professional library at a fraction of the cost of purchasing new books. And it can be enjoyable hunting through the shelves each month to see what new "treasures" might be found. There is nothing like having a room with floor-to-ceiling and wall-to-wall bookshelves full of books.

If analysts do purchase books to create a private collection, it is recommended that, where possible, they buy hard-bound editions. These will last longer because they are more durable, especially as the years pass. Some private, professional libraries have hard-bound intelligence-related books dating back to the First World War that was bought for as little as two dollars in second-hand shops. These texts have not been seen in any other special collections. The point is that books do not have to be the latest release to be of value—historical information is continuously valued in intelligence research (recall our discussion of "basic intelligence" in Chapter Two).

In the mid-1970s, the Central Intelligence Agency's Office of Security prepared a book outlining sources of information for federal investigators. The guide was entitled *Where's What*[54] and documented thousands of sources of information. Many of these are publicly available, but others can only be accessed by government employees with the appropriate authority. Even though the reference is now dated, the sources remain relevant today as they did then (though it is likely that the sources now appear under a different departmental name due to restructuring and corporate stylizing).

---

54. Harry J. Murphy, Office of Security, Central Intelligence Agency, *Where's What: Sources of Information for Federal Investigators* (New York: Quadrangle/The New York Times Book Co., 1975).

There are numerous sources of information in the form of professional and academic documents. For example, analysts can access conferences and symposia papers, data published by professional associations, as well as college and university professors and subject matter experts in industry.[55]

The United States has its government printer, the Library of Congress, *Congressional Record*, National Archives, and information available through the *Freedom of Information Act*. In most liberal democracies, there exist equivalents to the freedom of information legislation allowing access to the restricted holdings of a nation's archives.

Finally, public radio and television broadcasts can provide a vast range of information, as do photographic and motion picture archives. Many of these sources are now available in electronic form and can be downloaded in formats that can be played or viewed on the analyst's computer workstation.

Some examples of less-thought-about sources of open-source information can also include telephone directories (e.g., backdated); city directories; vehicle license plates; drivers' licenses; birth, death, and marriage records; civil and criminal court records; property titles, mortgage documents, liens, and caveats; school records; voter registration lists; credit reporting agencies; utility companies; credit card companies; insurance companies; stockbrokers; moving companies; chambers of commerce; racing or gaming commissions; banks and finance companies; the postal authority; numerous government departments, agencies, and statutory bodies (local, state/provincial, and national); and employment agencies.

---

55. Mark M. Lowenthal, *Intelligence: From Secrets to Policy*, second edition (Washington, DC: CQ Press, 2003), p. 79.

Like library materials, some of these sources are now available online, but for historical research, a trip to the library to view, say, an old telephone directory may be essential.[56] Although one might question this, consider a research project where an analyst is probing the *bona fides* of, perhaps, a walk-in defector to see if he is genuine, or whether his past was the result of fabrication through the creation of a cover, or *legend*.

Much of the material discussed so far finds its way to the analyst's desk via one or more conventional avenues for publishing. These publishing methods are usually associated with a commercial publishing house of some kind—usually a well-recognized publisher.

There is another avenue where information is presented to the public domain, which is termed gray literature in library sciences. It comprises information published in an informal manner that can vary in format—from print to electronic. The gray literature is usually associated with technical reports, scientific papers, working papers and notes, committee reports, white papers, and academic preprints.

These "publications" originate from various sources— for example, university and college research centers, think tanks, consulting firms, private researchers, government bodies, non-government organizations, and commercial businesses. There has been an exponential increase in these gray publications because of the ubiquity of digital technology.

Master's theses and doctoral dissertations are not considered part of the gray literature because, although they are prepared for a form of publication, they are readied for examiners, who perform a form of peer-review

---

56. e.g., As in a cold-case investigation.

of the research. Although the gray literature presents a rich source of information for analysts, it should be noted that these publications are distributed outside of the peer-review process and hence should be viewed with caution until each is assessed for validity and reliability.[57]

As a guide, there are three areas where analysts should direct their focus as they triage gray publications: (1) the nature of the document; (2) how the document is distributed; and (3) the source of the document. Regarding the latter, the source of funding for the research presented in the publication may influence the objectivity of the material.

Questions the analyst can ask of the material include: Where does the sponsor's funding come from? Has this been made transparent in the document? And has the funding been provided to benefit some organization in particular?

Users of gray literature praise this sometimes overlooked and underutilized source, but equally, users point out the limitations and stress the need to evaluate every document before placing any weight on its contents. It has been reported that some gray literature is counterfeit—that is, there are documents that are self-serving and masquerade as authoritative. These documents promote a particular point of view under the guise of being an authentic contribution to knowledge.

> The most common complaints against the gray literature—low quality and low accessibility—might appear to cancel each other out, but quality and accessibility are not correlated. Most reports contain some useful information, and the best are excellent. ...
> The big international NGOs (BINGOs) have more

---

57. See Chapter Six for more about evaluating intelligence information.

money than most local researchers in the tropics, so their reports may have useful information derived from remote sensing, camera-traps, and other resource-intensive techniques, while government departments have access to data that are not available to other researchers.[58]

Nevertheless, some sources promote bogus documents from industry groups, commercial firms, politically affiliated non-government agencies, politically sponsored think tanks, and others. This is not to say all these types of groups are involved in questionable publication practices. Far from it, nonetheless, users of the gray literature have reported that there are publications that present the pretence of academic rigor, but their publication is merely part of a media campaign to promote a political message or to increase a company's profit or organization's standing.

As a final thought about gray literature—it is a useful supplement to the peer-reviewed subject literature but should not be a substitute for it.

---

58. Richard T. Corlett, "Trouble with the Gray Literature," in *Biotropica*, Volume 43, Number 1, 2011, p. 3.

— CHAPTER SIX —

VALIDATING SOURCES

O pen-source information has been a boon for intelligence analysts. But it has also presented problems, particularly regarding testing a data item's authenticity.

In research terms, every analyst is concerned with information's *reliability* and *validity*.[59] Reliability refs to the method of collection will yield the same results with repeated applications. Validity refers to the accuracy of the information collected.

In addition to these considerations, some scholars point out that other axioms apply to intelligence information—relevance, legality, and security grading.[60] These three axioms will be discussed in Chapter Eight when we look at ethical considerations of open-source intelligence.

Reliability is the term researchers use to reflect a data item's trustworthiness. Other expressions often heard are credibility, honesty, and fidelity. Whatever term is used, it is important that the source can be trusted. A standard method of evaluating data is by using the Admiralty

---

59.   Gennaro F. Vito, Julie Kunselman, and Richard Tewksbury, *Introduction to Criminal Justice Research Methods: An Applied Approach, Second Edition* (Springfield, IL: Charles C. Thomas, 2008), pp. 66.

60.   Specifically, Marilyn B. Peterson, "Collating and Evaluating Data," in Marilyn B. Peterson, editor, *Intelligence 2000: Revising the Basic Elements* (Lawrenceville, NJ: International Association of Law Enforcement Intelligence Analysts, 2000), pp.91–94.

System. This procedure is also referred to as the NATO System, and "source and reliability matrix,"[61] which comprises a two-character code.

The first notation is an assessment of the level of confidence attributed to a piece of information. The second alphanumeric rating relates to its probability of being true. For instance, a piece of information classified as A-1 is considered *completely reliable* (i.e., A), and it has been *confirmed* (i.e., 1).

Table 2—Source reliability codes.

Admiralty Ratings			
Code	*Descriptors*	*Estimated Truth*	*Plus or Minus*
A	Completely Reliable	100%	0%
B	Usually Reliable	80%	10%
C	Fairly Reliable	60%	10%
D	Not Usually Reliable	40%	10%
E	Unreliable	20%	10%
F	Cannot be Judged	50%	10%
G	Unintentionally Misleading	0%	0%
H	Deliberately Deception	0%	0%

Table 2 shows an example of the first part of the NATO System—reliability—that the author has enhanced to include categories for *misinformation* (which is unintentionally incorrect information) and *disinformation* (which is outright deceptive information, that is,

---

61. U.S. Department of the Army, *Human Intelligence Collection Operations, Field Manual 2-22.3 (FM 34-52)* (Washington, DC: U.S. Army, 2006), B1-B2.

intentionally dishonest).[62]   Table 3 presents the method used to estimate a data item's accuracy.

As Prunckun[63] stated, evaluating information is an integral part of the analytic process. It usually takes place when the data are gathered. This information is evaluated according to the NATO System by asks questions such as:

- How reliable is the information source?;
- Has the source provided information before?;
- How accurate is the information?;
- What biases does the source have?;
- How recent is the information?; and
- How did the source access the information?

Table 3—Information accuracy estimates.

Admiralty Ratings			
Code	Descriptors	Estimated Probability of Truth	Plus or Minus
1	Confirmed	100%	0%
2	Probably True	80%	10%
3	Possibility True	60%	10%
4	Doubtful	40%	10%
5	Improbable	20%	10%
6	Cannot be Judged	50%	10%
7	Misinformation	0%	0%
8	Disinformation	0%	0%

62. Source, Hank Prunckun, *Methods of Inquiry for Intelligence Analysis, Third Edition* (Lanham, MD: Rowman & Littlefield, 2019), pp. 44–45.

63.   The following discussed was adapted from Prunckun, *Methods of Inquiry for Intelligence Analysis, Third Edition*, pp. 43–46.

Deception is a perennial concern with some types of intelligence, particularly national security intelligence and military intelligence.[64] In such cases, analysts need to evaluate data to distinguish between objective information and that tainted by bias. By way of example, take the accuracy code 6 in Table 2, misinformation; analysts should be mindful that they may obtain data that are *unintentionally* incorrect, illogical, or contradicted by other sources. In these cases, a code of 7 is appropriate. As for disinformation (code 8), these are data that are revealed (by other sources) to be *deliberately* false or misleading (i.e., provided for deception, perhaps as part of an opposition's counterintelligence operation).

Secondary sources, such as government press offices, commercial news organizations, non-government organization spokespersons, and other information providers, can intentionally or unintentionally add, delete, modify, or otherwise filter the information they release to the public. These sources may also convey a message in English for North-American or international consumption, and a different non-English message for local or regional consumption. It is important to know the background of open sources and the purpose of the public information to distinguish objective facts from information that lacks merit, contains bias, or is part of an effort to deceive the reader.[65]

---

64. Michael Taylor, "Open -Source Intelligence Doctrine," in *Military Intelligence*, Volume 31, Number 4, October–December 2005, p. 14.

65. U.S. Department of the Army, *FMI 2-22-9: Open-Source Intelligence* (Washington, DC: U.S. Army, 2006), p. 2–10.

Ideally, the evaluation process is applied to each piece of information collected.[66] However, in agencies that manage large volumes of data, this may be an automated process where a generic rating is assigned if the data are merely stored. Still, when it is retrieved for use, the data item is then reevaluated.

How would this work in practice? Imagine an analyst obtains a piece of information from a person posting a comment on a blog. On the one hand, if this person is a new source that has never been exploited before, the reliability of this piece of information would therefore have to be F—the reliability cannot be judged. On the other hand, if the information obtained came from a person that has been the source of previous information and has proven to be truthful in almost every instance, then an accuracy code of 2 would be assigned. In either case, the two-character code would be printed on the document to show its overall rating.

Customarily, accuracy rating precedes the reliability code—for instance, F-2. Having said that, it needs to be pointed out that the ratings must be logical; assigning a rating of, say, E-2 (unreliable source but probably true) or H-3 (deliberately deceptive but possibly true) would raise questions in the consumer's mind about whether the evaluation process was rational.

Although the NATO System represents an objective position, ratings are derived through a subjective process because judgment plays the role. When assigning a code, especially for cyber sources, the analyst should cross-check the data against an established valid source when considering accuracy.

---

66. Jack Morris, *Police Intelligence Files: An Introduction to the Use of Confidential Police Files* (Orangevale, CA: Palmer Enterprises, 1983), pp. II-8–II-11.

Regarding credibility, the analyst should check the source's URL, the location and ownership of the server, the domain and/or platform's appropriateness for the material presented, and what other Web sites link to the target source and if the target has been reviewed as reliable.[67]

Evaluation is a complicated process but an important one, as personal or agency bias can adversely affect the results of an intelligence research project. A good evaluation results from the source's reliability being assessed independently of the information's credibility.

67. Robert M. Clark and William L. Mitchell, *Deception: Counterdeception and Counterintelligence* (Los Angeles: Sage, 2019), p. 177.

— CHAPTER SEVEN —

ANALYSIS

I ntelligence is secret research. It involves collecting information that, once analyzed, provides insight into matters of import. Simply put, intelligence reduces uncertainty in decision making.[68]

In this chapter, we will discuss how open-source information can be analyzed. But first, we need to examine the type of data we have to work with—is it unstructured or structured data? If it is the former, it is referred to as *qualitative,* and the latter, *quantitative.*

Each type of data has analytical methods suited to each. And, there are many techniques for each of the categories. In fact, there are entire textbooks devoted to qualitative, quantitative analyses.

Because this chapter intends to introduce you to the analysis of open-source information, it will limit itself to a high-level view of the topic.

If we look at qualitative data, we see that this is perhaps the most common form available because it is so pervasive. It takes the form of, in most cases, words and concepts (e.g., *themes* found in word expressions). Although, qualitative information also includes photographs, drawings, illustrations, videos, maps, and many other forms of imagery. Quantitative data involves numbers.

---

68. Note that intelligence does not eliminate uncertainty.

To better understand qualitative and quantitative information, we will start our examination by looking at four types of information. An understanding of this typology will enable us to determine the analytic techniques best suited to each.

The four types of data are nominal,[69] ordinal, interval, and ratio. Nominal information is the type that can be placed into two or more categories, but there is no underlying order to these data. Nominal information is the lowest level of measurement of the four and can be considered qualitative. For instance, "yes" or "no"; "present" or "absent"; "male," "female," or "non-binary"; and so on. This type of data can come from information that has been collated—such as that from a bureau of statistics—or from unstructured (word-based) sources—such as transcripts of interviews, news reports, blog posts, etc. (i.e., by providing a structure for this unstructured information), or numerical data by "reducing" the data measurement level to this simple scale.

Ordinal data shares some similarities to nominal information in that themes are placed into categories, but in addition, these categories have an order. For example, economic status can have several groups, but also an order—low, medium, and high. Take, as an example, the aggression level of several outlaw motorcycle gangs. They could be categorized[70] as "very aggressive," "somewhat aggressive," "neutral," somewhat unaggressive," and "very unaggressive."

These ordinal categories can be derived from reading police incident reports (i.e., narrative descriptions) or by

---

69. Along with ordinal data, these two types are sometimes referred to as *categorical* data.

70. Of course, the number of categories can be increased or decreased, and alternative descriptors used.

counting, the number of acts of violence attributed to each gang from statistical information. If done via a conversion from numeric data, the process is an example of reducing higher-level data to a lower scale.

Interval data is like ordinal data (i.e., these provide classification and order), except that the intervals between the categories are equally spaced. For instance, take the measurement of temperature—either Fahrenheit or Celsius. Both are thermometric scales that measure the points where water freezes and boils. With the former, these temperatures are 32° and 212,° respectively. On the Celsius scale, these temperatures are 0° and 100°. Although the spacing between each degree is uniform, there is no point on either scale where there is no energy whatsoever (unlike the Kelvin scale). In this regard, interval data is numeric.

Ratio data is related, metaphorically, to interval data because it too is for numerical information. Like interval data, ratio data allows for differences between values, but rather than having arbitrarily assigned zero-point, ratio data has a natural zero.

Suppose we imagine two illicit drug labs—one that can produce 10 kilograms of methamphetamine during a fixed period and another that can produce 20 kilograms in the same time frame. In that case, we can state that the former lab has half the capacity of the second. This is because the interval scale has a natural zero point represented by no weight at all. Therefore, 20 kilograms is twice as much as 10 kilograms.

Why does it matter whether a data item is nominal, ordinal, or interval, or ration? Because knowing what scale a piece of information is related to allows the most appropriate analytic technique to be selected. Selecting the most appropriate method is vital because some methods only work with certain types of data. For instance, some techniques will work with categorical data (i.e., nominal, or ordinal data), while others will only work with numerical data (i.e., interval, or ratio data). Yet, others will work with a mix. Table 4 shows the types of operations that can be performed on each data type.

These measurement levels specify the precision of the data. Why does this matter? Because as the precision level increases, it represents more complexity in the measurement. The more complex the measurements, the more analytical techniques can be performed, such as statistical tests.

Table 4—Data Types by their attributes.

	Nominal	Ordinal	Interval	Ratio
Categories	✓	✓	✓	✓
Rank order		✓	✓	✓
Equal spacing			✓	✓
True zero				✓

Does this mean that nominal data cannot be subjected to statistical tests? No! One of the most straightforward statistical tests—the *chi-square*—can be run on nominal data. If we recall that nominal data can be derived from qualitative information by nothing more than collation, the possibility of running a statistical test presents a powerful way of analyzing the same information.

As previously mentioned, data measured by a higher scale—such as ratio data—can be reduced to a lower scale measurement. Hence, these data can be analyzed using more advanced methods. But, lower-level data cannot be converted into a higher scale (e.g., nominal data comprising, for instance, "selling drugs" / "not selling drugs" categories cannot be made into ordinal, interval, or ratio items). Table 5 summarizes the types of operations that can be performed on each data type.

Table 5—Data types and operations.

Operations	Nominal	Ordinal	Interval	Ratio
Equality[71]	✓	✓	✓	✓
Order		✓	✓	✓
Add & subtract			✓	✓
Multiply & divide				✓
Mode	✓	✓	✓	✓
Median		✓	✓	✓
Mean			✓	✓

What does this mean for how data is analyzed? As a guide, if we are concerned about *probability*,[72] and the data are nominal or ordinal, then non-parametric analyses can be used. If the data are interval or ratio, then parametric tests can be used.

---

71. In this context, the term *equality* refers to whether the data item's fits into a category (=) or not (≠).

72. Probability is a numerical expression of how likely an event is to occur, or the likelihood that a proposition is true. One such scale is that over 80 percent indicates that issues under study is *most likely* to occur. Between 60 and 80 percent it is somewhat likely to occur. Between 40 and 60 percent it is an even chance that it will occur. Between 20 and 40 percent it is somewhat unlikely. And, below 20 percent it is highly unlikely to occur. Peterson, *Applications in Criminal Intelligence: A Source Book*, p. 51.

What is the difference? Parametric analytical tests rely on the data having normal distribution, whereas non-parametric procedures do not (i.e., a skewed distribution). The decision to use over the other depends on whether the mean or median more accurately represents the center of your data set's distribution. For data where the mean represents a normal distribution (and the sample size is large enough), then a parametric test is appropriate. For data sets where the median reflects the distribution's center, then a non-parametric test is suitable.

Nevertheless, non-parametric tests apply to a broader range of problems. For instant, the most straightforward test to apply is the chi-square. This test allows the researcher to determine whether the distribution in, say, a nominal dataset has occurred by chance (i.e., random variation), or was the result of another factor (i.e., a *variable*) acting on it. This is expressed in a p-value, ranging from 0 (impossible) to 1 (almost certain).

Table 6 shows the p numeric range with corresponding qualitative descriptors. The astute observer will note that the p-values correspond to percentages—0%, 25%, 50%, 75%, and 100%.

Table 6—p-value descriptors.

p-Value	Descriptor
0	Impossible
.25	Unlikely
.5	Possible
.75	Likely
1	Almost Certain

Rather than taking your various pieces of information and subjecting them to an "expert opinion" based analysis method, using your understanding of data types affords you the ability to apply a well-defined technique that allows for *transparency* and *replication*.

Table 7—Examples of qualitative analyses.

Name	Description
Association/link/network analysis	Diagrammatically shows the relationships between people, organizations, events, etc.
Descriptive analysis	A factual summary of a person's activities, an event, a group, etc.
SWOT analysis	Evaluates strengths, weaknesses, opportunities, and threats as a way of planning how to reach a tactical objective or strategic end-state.
PEST analysis	Evaluates the political, economic, social, and technological factors of the target's environment.
Pros-and-Cons	Weigh's-up the factors in favor of action with those against.
Forcefield analysis	Similar to a pros-and-cons analysis, but assigns a quantitative score to indicate the magnitude of the issue that may be either "driving" or "restraining."

Transparency and replication are at the heart of the scientific method of inquiry. These two attributes allow peers to scrutinize what you have done and, in theory, allow them to take the same data and subject it to the same analyses to arrive at the same conclusions.

You may say, "I work with quantitative information; some of this discussion seems irrelevant to me." If you work with qualitative data, then knowing these distinctions will be invaluable to your analysis. How? In the next chapter, we will explore a few illustrative reports demonstrating how qualitative analytic techniques cited in the list below can be used.

— CHAPTER EIGHT —

REPORTS

I n Chapter Four, we spoke about operational assessments. We also mentioned target profiles, biographical sketches, chronologies, and other analyses. Here, we will summarize these reports to give analysts an idea of what they can offer an investigative team.

OPERATIONAL ASSESSMENTS

First, let's look at the take the task of writing an operational assessment. Recapping, the purpose of an operational assessment is to examine issues that are "...broader than what is happening at the tactical level..." Although operational assessments fall short of what would be expected in a strategic assessment, the objective of this type of report is to move from being reactive—as with tactical reports—to be anticipatory.

The operational assessment looks at crime problems in a broad way. "The value of [operational assessments] to law enforcement administrators beyond the immediate arrest and prosecution can be significant in determining how, where, and with what degree of intensity resources should be allocated."[73]

What does an operational assessment look like? How is it laid out? And, how many pages or words is it? These are reasonable questions that are not usually answered by

---

73. Justin J. Dintino and Frederick T. Martens, *Police Intelligence Systems in Crime Control* (Springfield, IL: Charles C. Thomas, 1983), p. 114.

books on intelligence analysis. Perhaps, the authors anticipate that these reports are being "taught" in-house by the analyst's agency? Regardless, it would be beneficial to see an outline of topic headings and what should appear in each of these sections.[74]

Introduction

Provides brief details about the legal basis or agency policy that provides the analyst authority to compile the report, and the name of the authorizing officer (e.g., the officer-in-charge of the investigation or the intelligence unit's manager, etc.), date, and file number for audit purposes. These details can be set out at the top of the page like a memorandum.

Background

The background section contains a statement of the assessment's objective as well as a description of how the problem or issue under investigation arose. The length is usually one or two paragraphs.

Aim

The aim is the report's "research question." The aim acts to guide the study and keep the inquiry focused on a specific issue (as well as what is in scope and what is not). This helps keep the investigation from "wandering" into areas that could be interesting but not relevant to the outcome of the issue being perused.

---

74. The following discussion is based on material presented by the author is his book, Pruckun, *Methods of Inquiry for Intelligence Analysis, Third Edition*, pp. 153–163. See this reference for more details, as well as an example of an operational assessment.

Current Situation

This section may comprise one section or several subsections of descriptive data about the problem. These data can come from several open sources.

In addition to describing the phenomenon, the analyst can explain the ramifications of the problem in its historical, social, economic, political, religious, cultural, or anthropological context, and its extent in the jurisdiction (e.g., the region, the nation, or around the globe). But it is important not to interpret the information here in this section—that comes after analysis, in the prognosis section.

The analyst could explain any progress being made, pitfalls encountered during the investigation, or the implementation of interventions to date. This information sets the scene for the next section, which is the analysis.

The length of this section might be between three or four paragraphs to a page-and-a-half. Recall, the heading is "current situation," so only information about the current situation should appear here. Do not repeat details mentioned in other sections.

Analysis

This section is where the analyst presents the results of their analysis using techniques such as statistical analysis, network analysis, force field analysis, SWOT, PESTO, or others depending on the type of data being used.

Only analysis appears in this section (e.g., narrative, tables, or figures[75]). Do not repeat details mentioned in other sections. The length should be kept to a page or less.

---

75. "Graphs, charts, photographs, drawings, diagrams, as well as other terms are all considered *figures* when writing an intelligence report. The term *figure* is a distillate of all

Prognosis

The prognosis section is essentially a discussion of the analysis. Still, it extrapolates from the findings to assess what is revealed about the activity under investigation and what can be done to provide relief. The analyst may consider a change in focus from what was being done to a modified or new approach, or a shift in priorities using the same interventions, etc.

It is common to present this discussion within the frame of the results of the analysis. Although framed in the logical order of the analysis, this is an exercise in deductive reasoning in a narrative form—or "thinking out loud" about each of the issues discovered in the analytic process that preceded this discussion.

To produce a range of options for decision-makers to consider, this section can talk about the agency's strategic mission/goals or key performance indicators (KPIs) and how possible interventions may impact these benchmarks. Decision-makers around the conference table can then argue priorities and resources according to "what works," "best value," or "best practice." Issues that might be discussed in the prognosis section could be generated from any one (or more) of the topics contained in the *five Is model*:

o Intelligence. Issues relating to information gathering, collation, and analyzing (past or future);

o Intervention. Tactics to block, disrupt, weaken, or eliminate "the problem";

---

graphical representations into one summary term. The only other term that is used in presenting research results is the term *table*. *Figures* and *tables* are the only two terms that should be used in intelligence writing." Prunckun, *Methods of Inquire for Intelligence Analysis, Third Edition*, p. 126.

- o Implementation. Translating the goal of the proposed intervention (theory or principles) into practical methods in the field;

- o Involvement. Ways to get other agencies (or companies, organizations, and individuals) to contribute somehow to being part of the implementation of the intervention(s); and

- o Impact. How the problem will be evaluated and by whom. The evaluation may be simple or complex, but because the problem is one of an operational nature, a basic evaluation is most likely all that is needed (i.e., an output-based evaluation rather than one that is outcome focused).

Recommendations

The officer-in-charge of the investigation or the intelligence manager dealing with operational issues that are within the scope of this type of assessment will require options for consideration. Leveraging the discussion in the previous section, the analyst needs only restate the range of options available.

This can appear in the form of a bulleted list to simplify what is possible. If there is a preferred option, this can be highlighted in some way—for instance, appearing first in the list with the other options appearing in a list below in diminishing order, with the least preferred at the end.

To help frame a set of recommendations, use the strawman technique discussed in chapter 6. This technique allows decision-makers to understand the strengths/benefits of the preferred option when contrasted with other options.

Appendices

Because an operational assessment is a concise report, any attachments that may be included need to be kept to a minimum. Examples of attachments might include a map, a photograph, or an organizational chart showing relationships of the targets rather than in a drawn-out narrative.

TARGET PROFILES

A *target profile* is a type of intelligence report that summarizes information about a specific target. It is an operationally-focused description that is considered a short-form report. Joby Warrick, in his book on a CIA double agent operation, describes the role a target profile plays: "Like an artist assembling a giant mosaic, [the targeter can] summon bits of information from wiretaps, cell phone intercepts, surveillance videos, informant reports, and even news accounts [as well as other open sources] . . . to develop a profile that the agency's spies, drone operators, and undercover case officers [can use operationally]."[76]

A target can be an individual or group, but it could also be a company or organization. In the latter case, the report may be titled a *criminal business profile, terrorist organization profile*, or similar name. A variation to the target profile is the *problem profile*—this is a report that focuses on an issue, not a person or organization. An example could be a series of crimes occurring in a geographic "hot spot." In any case, the target can be either the subject of a current inquiry or an emerging target for a proposed investigation. The report summarizes what is known about the target and, in doing so, identifies information gaps, which feed into a collection plan for additional data.

Target profiles often provide investigators and field operatives with a range of options regarding possible

---

76. Joby Warrick, *The Triple-Agent: The al-Qaeda Mole that Infiltrated the CIA* (New York: Doubleday, 2011), p. 69.

intents (i.e., hypothesis derived from inductive reasoning) or potential ramifications if the target continues the activity at the center of the report's concerns (e.g., risk assessment). In doing so, a good report will prioritize the need to make further inquiries about the target in ranked order with other targets under investigation so that intelligence unit managers can allocate resources.

A target profile consists of several sections that are arranged to "tell a story":[77] (1) background, (2) personal details, (3) business details, (4) criminal record (in the case of a law enforcement target), (5) criminal associates, (6) physical environment, (7) analysis, and (8) target planning (i.e., Having presented the facts and analyzed these to gain an understanding into the target's activities, the target planning section takes the insights developed, and answers the question of "where to from here?" or "so what do we do now?"). There may also be attachments appended to the end of the report, but these data need to be informative, not "padding."

BIOGRAPHICAL SKETCHES

A specific type of tactical profile is the *biographical sketch*. This report provides a briefer version of the tactical profile and presents without analysis—just the facts. A biographical sketch outlines a target, person-of-interest, or an associate of either. These vignettes can also be compiled on businesses, organizations, or groups. The report's layout needs to be easy to read, like an entry in a *Who's Who* directory, though the form can vary to suit the reader's preference.

For individuals, information items could include, but are not limited to, the person's name, nickname(s), aliases,

---

77. For a fuller discussion and a practical example, see Prunckun, *Methods of Analysis for Intelligence Analysis, Third Edition*, pp. 139–152.

current and former addresses, occupation and current and former employers, hobbies, places travelled, friends and associates, and other items published in news articles or social media, including photographs illustrate behaviours or events.

Not all these details may be available via open sources, so once exhausting this avenue, official records and confidential sources can be added and/or compared to the data open-source information for inconsistencies.

A business sketch might record its incorporated name, government registration number, and the names and details of the directors, secretary, and public officer. Details about the company's share issue and net worth, as well as other facts gleaned from the public register of businesses. There are, of course, many more facts that can be sourced from the business's Website and the variety of open-sourced discussed in previous chapters.

CHRONOLOGIES

Chronologies can be presented as either tables or timelines. Although timelines are often easier to draw and understand because of their simple visual form, chronological tables can be helpful in other situations.

These elementary forms of analysis show a time-based representation of events. They can also indicate future events when used as a planning aid. Many word-processing and spreadsheet programs have templates for drafting timelines. These templates can be used as-is or modified to suit a departmental style.

CRIME BULLETINS

Producing a *crime bulletin* is an efficient way for analysts to disseminate newly obtained information or alert others about a developing issue concerning the matter under investigation. As with most analytic products, the form of this report is not critical, although the precision in its brevity is important.

Depending on the issue being raised, crime bulletins are usually a page or less that comprise a contextual summary, a statement of the problem, and the new information or alert (e.g., BOLO—be on the lookout). Whatever information is needed to convey the message can be included, but no more—a short narrative, photograph, map, or drawing.

DEMOGRAPHIC ANALYSIS

An analysis of the demographics of a community, group, gang, or place can be a stand-alone report, or it can be a section within, say, an operational assessment. The report's focus is to summarize, as in a case study, factors such as age, sex, race, ethnicity, education, income, and family status.

These data are readily available for a national bureau of statistics. The idea is not to write a precis of what can be found in government statical tables, but to set the crime problem in the context of the "picture" these data present.

In the social sciences, researchers use demographic analyses to inform decision-makers about emerging problems. Criminologists have adapted this method to forecast crime and identify illicit drug markets or areas where certain offences are expected to be more than others.

For instance, demographic research has shown that males between the ages of eighteen and twenty-five are more prone to behaving violently.[78] Therefore, knowing the location of such a cohort, an understanding of the

---

78. Of course, there are other social, economic, and political variables that need to be considered, but this an illustrative example only.

geographic area can assist crime prevention officers design interventions. In effect, as a stand-alone approach, demographic analysis fits with *intelligence-led policing*[79]—analysis social conditions that contribute to crime.

DESCRIPTIVE ANALYSIS

In some cases, information may be best presented in diagrammatic form—figures, charts, graphs, and activity maps—because to describe the data in narrative form would be somewhat overwhelming to the reader. However, in other situations, diagrams are not able to explain circumstances well. As such, a descriptive analysis might be the best way.

These reports summarize an event or an activity, or a person, group, or organization. Think of a well-written newspaper article where the reporter "tells the story" by explaining the development or problem (i.e., the context) and then describes the facts surrounding it.

At the end of the description, an inference is drawn from these data[80] and a commentary on what this reasoning might mean for the operation underway, or later, for future legislation/policy. By way of example, a report of this type lends itself to reporting on, say, a group of organized criminals that may have had a change in the

---

79. Intelligence-Led Policing is associated with *problem-oriented policing* because the two methods support the broader policing paradigm of *community-oriented policing*. Herman Goldstein, *Problem-Orientated Policing* (New York: McGraw-Hill, 1990).

80. Although conclusions are contained in the data, inductive reasoning does not exclude alternatives, as would be the case with deductive reasoning. Noel Hendrickson, *Reasoning for Intelligence Analysis* (Lanham MD: Rowman & Littlefield, 2018), p. 18.

group's leadership, territory, or illicit business model, mention just a few issues.

NET WORTH ANALYSIS

An analyst does not need to be a financial investigator to perform this basic analysis because it is a high-level examination of a person or group's financial position. This type of analysis can be performed as a stand-alone report or as a section within a target profile or operational assessment to determine whether the target is living within the means of their reported income.

If the analogy of a paramedic and a surgeon is used, the intelligence analyst would be the paramedic performing the first stage analysis while calling in the surgeon to perform the more intricate analysis once the issues are identified.

Net worth is an indirect method of assessing the target's income. Therefore, it is a helpful method in situations where the analyst may have come across information that suggests some form of illegal enterprise. It could also be used to assess a target's suitability to an approach by a field operative hoping to recruit the target as an agent (e.g., an offer of financial assistance).

Net worth is simply the difference between the target's assets and liabilities. If the analyst conducts a net worth analysis over a period, say, for the end of each financial year, they can compile a picture as to whether the target is growing in worth or is experiencing losses and what the magnitude of these gains or losses might be.

The formulas for calculating net worth are as follows, step by step:

1. Assets—liabilities = net worth
2. Net worth—prior year's net worth = increase or decrease in net worth

3. Net worth increase (or decrease) + living expenses = income

4. Income—funds in known sources = funds from potentially illegal sources[81]

---

81. Leigh Edwards Somers, *Economic Crimes: Investigating Principles and Techniques* (New York: Clark Boardman Company, 1984), p. 99.

— CHAPTER NINE —

ETHICAL CONSIDERATIONS

D iscussions about morality, politicization, guideposts, principles, codes, creeds, and values are not the expected reading material for an intelligence analyst. Indeed, such a collection of ethics-based issues is an unlikely feature of intelligence tradecraft. Yet, intelligence analysts face the dilemma of acting ethically while at the same time engaging in what some have portrayed as an unethical business—spying.

Bernard Newman once wrote, "A spy must be a [person] of integrity and yet must be prepared to be a criminal."[82] Whether the spy is a field operative or a "scholar-spy,"[83] there are ethical issues that need to be considered.

However, we are discussing open-source information, so are there ethical constraints associated with publicly available information? Are there legal issues? The short answer is "yes" … and "no." Let us look at a few situations.[84]

---

82. Bernard Newman, cited in John Alfred Atkins, *The British Spy Novel: Styles in Treachery* (London: John Calder, 1984), pp. 142–143.

83. The term *scholar-spy* refers to professors, academics, and other researchers who are engaged in intelligence analysis.

84. The following discussion was adapted from Hank Prunckun, *Handbook of Scientific Methods of Inquiry for Intelligence Analysis* (Lanham, MD: Scarecrow Press, 2010), pp. 212–215.

There are unlikely to be any ethical issues when it comes to national security intelligence—intelligence agencies that target foreign threat actors.   Still, for agencies that target domestic national security threats, there are concerns.   These agencies typically include federal police forces and other state/provincial and local law enforcement departments.

Why would this be the case?   Because there are domestic laws that safeguard individuals' rights.   Some time ago, the International Association of Chiefs of Police (IACP) put forward a set of guidelines designed to protect the rights of individuals genuinely not involved in criminal acts.

In the main, these guidelines urge that when an individual's activities are not criminal, the information should not be recorded.   Likewise, in the case of organizations, unless an organization's "ideology advocates criminal conduct and its members have planned, threatened, attempted or performed such criminal conduct,"[85] it is both unnecessary and wrong to gather such information, whether from surreptitious means or via open sources.

Moreover, information about an individual's "sexual, political or religious activities, beliefs or opinions, or any dimension of private life-style" should not be collected or recorded in intelligence files unless that information is material to a criminal investigation.[86] This advice applies to domestic law enforcement intelligence but does not apply to foreign national security intelligence.   In the latter

---

85. Los Angeles Police Department, *Standards and Procedures for the Anti-Terrorist Division* (Los Angeles: Los Angeles Police Department, 1984), p. 2.

86. Los Angeles Police Department, *Standards and Procedures for the Anti-Terrorist Division*, p. 2.

case, these data would be necessary for compiling a psychoanalytic profile of a target or producing another specialist report.

Moreover, when evaluating the source of publicly available information, if the source proves unreliable, that data should not be stored in criminal intelligence files. Likewise, such information collection should be strictly limited unless it passes the "sunlight test."[87] Therefore, a set of standards is necessary to create an ethical law enforcement intelligence filing system. The elements of such a system include:

1. Specific guidelines for determining:
   a. the kind of information that should be kept in intelligence files;
   b. the method of reviewing the material as to its usefulness and relevance; and
   c. the method of disposing of material purged from intelligence files considered to be no longer useful or relevant.
2. Systematic flow of pertinent and reliable information.
3. Uniform procedures for evaluating and validating information.
4. A system for proper analysis of information.
5. A system capable of rapid and efficient retrieval of all information.
6. Explicit guidelines for disseminating information from the files.
7. Security procedures.[88]

---

87. The *sunlight test* is a simple ethical assessment where a person asks themself, "Would I do this if what I do is printed on the front page of tomorrow's newspaper"?

88. International Association of Chiefs of Police, *Law Enforcement Policy on the Management of Criminal Intelligence: A Manual for Police Executives* (Gaithersburg, MD: IACP, 1985), p. 9.

Systematic purging of files[89] according to the guidelines alluded to above should ensure that information being retained (i.e., as per an intelligence collection plan) is related to legally authorized projects that are necessary to meet the decision maker's intelligence requirements.

Situations where data have become irrelevant because of their age should not be allowed to accumulate. For instance, a Rand Corporation study once found an American police department retained intelligence files that "contained information on suspected Nazis, the concern of the 1940s; Communist Party membership rosters, the concern of the 1950s; black militants, right-wing extremists, and anti-Vietnam war demonstrators, the concern of the 1960s."[90]

Intelligence managers in the twenty-first century would be hard-pressed to justify the retention of these types of data. Analysts should always be mindful of their targets. No matter how unpalatable their cause, analysts should be conscious of the difference between legal, lawful protest and subversion and criminal activity—distinctions not always correctly made in the past.

For the most part, secrets in the context of this book are secrets of the state (or a private corporation), and as such, keeping them secret will most likely be an obligation of a legal statute or a legally binding agreement. Because of this, and unless disclosure is covered by a countervailing legal instrument (e.g., a whistle-blower-type act, an order

---

89. Information discovered to be irrelevant to the issue under investigation should be disposed of according to the analyst's agency document destruction policy (e.g., shredding).

90. Brian Michael Jenkins, Sorrel Wildhorn, and Marvin Lavin, *Intelligence Constraints of the 1970s and Domestic Terrorism: Executive Summary* (Santa Monica, CA: Rand, 1982), p. 5.

of a court, a police warrant, a direction given by a judge or magistrate), then disclosure should not be made. However, if the secret is the cornerstone of an illegal activity, a cover-up, or it relates to the planning or commission of an unlawful act (or covered under a whistle-blower-type law), then revealing the secret is not likely to fall into this category. In fact, there may be a legal obligation to disclose these facts to a law enforcement or regulatory agency, or a civil rights or oversight body.

By way of example, take the U.S. Army's field manual *Open-Source Intelligence* that lists several intelligence-gathering activities that are prohibited. The banned activities are the improper collection, retention, or dissemination of U.S. person information:

- Gathering information about U.S. domestic groups not connected with a foreign power or international terrorism.
- Producing and disseminating intelligence threat assessments containing U.S. person information without a clear explanation of the intelligence purpose for which the information was collected.
- Incorporating U.S. person criminal information into an intelligence product without determining if identifying the person is appropriate.
- Storing operations and command traffic about U.S. persons in intelligence files merely because the information was transmitted on a classified system.
- Collecting U.S. person information from open sources without a logical connection to the unit's mission or correlation to a validated collection requirement.
- Identifying a U.S. person by name in an intelligence information report without a requirement to do so.

- Including the identity of a U.S. person in a contact report when that person is not directly involved with the operation.[91]

Finally, we need to look at business intelligence. There are two ethical hazards—the temptation to engage in some form of industrial espionage, but paradoxically, the determination not to engage in espionage may jeopardize the ethical gathering of information.[92]

At this point in our discussion, it should be evident that illegal data collection is not necessary to have an effective business analysis program. All information needed to carry out these types of analyses needed to guide a business can be obtained from open sources.

Yet, there is the problem where a business is so determined not to be perceived in unethical information gathering that it confuses legitimate methods with "spying." Take, for example, a case where a business asks one of its distributors about a competing firm's buying practices. Yes, the competitor may be disturbed to find out questions were asked, but this type of gathering is argued to be ethical.[93]

What must be avoided is the resort to bribery, electronic eavesdropping, or fraud to find out competitive information. Likewise, any attempt to exchange price or market share information with competitors could be a breach of anti-competition or anti-trust laws.

---

91. U.S. Department of the Army, *FMI 2-22.9: Open-Source Intelligence* (Fort Huachuca, AZ: Department of the Army, 2006).

92. John M. Kelly, *How to Check Out Your Competition* (New York: John Wiley, 1987), p. 68.

93. Kelly, *How to Check Out Your Competition*, p. 68

ABOUT THE AUTHOR

Dr Henry (Hank) Prunckun, BSc, MSocSc, MPhil, PhD, is an Adjunct Associate Research Professor with the Australian Graduate School of Policing and Security, Charles Sturt University, Sydney. He is a former Australian government intelligence analyst who spent much of his twenty-eight-year operational career in tactical intelligence and strategic research, but also served in security, investigation, and counterterrorism. He was conferred with two literary awards and a professional service award by the International Association of Law Enforcement Intelligence Analysts. After leaving government service, he worked initially as a freelance private investigator, then spent over a decade as a research criminologist studying transnational crime—espionage, terrorism, drugs and arms trafficking, and cyber-crime.

INDEX

Admiralty Code, 2

Admiralty Grading
System, 2

Amnesty International, 8

anti-competition laws, 69

anti-trust laws, 69

biographical sketches, 28,
52, 58–59

Bowen, Russell J., 31

business intelligence, 14,
17, 20–21, 28, 69

chi-square, 47, 49

chronologies, 59

community-oriented
policing, 61nf79

crime bulletins, 59–60

data: nominal, 45, 46–47,
48, 49; ordinal, 45–46,
47, 48; interval, 45, 46–
47, 48; ratio, 46, 47, 48

demographic analysis, 60–
61

descriptive analysis, 61

disinformation, 39, 40, 41

doctoral dissertations, 27,
35

ethical issues, 64–69

Foreign Broadcast
Information Service,4,
4fn9, 5, 18, 18fn37

gray literature, 35–37

Human Rights Watch, 8

information: reliability, 36,
38; typology, 45–51;
validity, 36, 38

intelligence community,
16–22

intelligence: first
principles, 28; typology,
16–21

intelligence-led policing,
61, 61fn79

Library of Congress, 14,
18fn36, 34

master's theses, 27, 35

misinformation, 39, 40, 41

NATO System, 2, 39–40,
42

net-worth analysis, 62–63

open-source intelligence:
collection, 30–37; ethics,
64–69; history, 3–9

operational assessments,
25, 29, 52–56

OSINT. *See* open-source
intelligence

peer-review, 35–36, 37, 50

problem-oriented policing,
61nf79

research questions, 1, 23–
25, 23fn44, 26, 28, 53

scholar-spies, 32, 64,
64fn83

target profiles, 28, 57–58

71

~ NOTES ~

~ NOTES ~

~ NOTES ~

~ NOTES ~

www.ingramcontent.com/pod-product-compliance
Lightning Source LLC
Chambersburg PA
CBHW022342280326
41934CB00006B/739